Grade 3 • Unit 1

Forces Around Us

FRONT COVER: (t)FreedomMaster/iStock/Getty Images, (b)ShutterStockStudio/Shutterstock;
BACK COVER: FreedomMaster/iStock/Getty Images; **SPINE** FreedomMaster/iStock/Getty Images

Mheducation.com/prek-12

STEM McGraw-Hill is committed to providing instructional materials in Science, Technology, Engineering, and Mathematics (STEM) that give all students a solid foundation, one that prepares them for college and careers in the 21st century.

Send all inquiries to:
McGraw-Hill Education
8787 Orion Place
Columbus, OH 43240

ISBN: 978-0-07-699626-1
MHID: 0-07-699626-3

Printed in the United States of America.

4 5 6 7 8 9 10 11 LWI 26 25 24 23 22 21 20

Table of Contents
Unit 1: Forces Around Us

Forces and Motion

Electricity and Magnetism

Forces and Motion

ENCOUNTER
THE PHENOMENON

What did the skateboarder have to do to get to the top of the ramp?

GO ONLINE
Check out *Skateboarders* to see the phenomenon in action.

Talk About It

Look at the photo and watch the video of the skateboarders. What questions do you have about the phenomenon? Talk about your observations with a partner.

Did You Know?

The very first skateboards had handles and were developed in California.

STEM Module Project Launch
Engineering Challenge

Design a Skatepark

You have been hired as an architectural designer. At the end of this module, you will develop a design for a skatepark. Your goal will be to design, build, and test a model that is able to get a marble from one side of the park to the other.

Lesson 1
Motion

Architectural designers apply their knowledge of motion, forces, and design to create playgrounds and skateparks.

Lesson 2
Forces Can Change Motion

What do you think you need to know before you can design a skatepark?

SAM
Architectural Drafter

STEM Module Project

Plan and Complete the Engineering Challenge Use what you learn throughout the module to complete the challenge.

LESSON 1 LAUNCH

How Does It Move?

Three friends were playing soccer. They each had a different idea about the ball's motion. This is what they said:

Desmond: *The way I kick the ball determines how far the ball will go.*

Aliyah: *The way I kick the ball determines where it will go and how far.*

Megan: *It doesn't matter how I kick the ball; it will move where the ball wants to go.*

Who do you agree with most? ______________________________

Explain why you agree.

____

____

____

____

You will revisit the Page Keeley Science Probe later in the lesson.

LESSON 1

Motion

ENCOUNTER
THE PHENOMENON

Why does the ride move like that?

GO ONLINE

Check out *Carnival* to see the phenomenon in action.

Talk About It

Look at the photo and watch the video of the carnival in action. What questions do you have about the phenomenon? Talk about your questions and observations with a partner. Record or illustrate your thoughts below.

Did You Know?

The California State Fair is set up in only four days and has more than 600,000 visitors.

INQUIRY ACTIVITY

Hands On

Moving Marbles

In the video, you observed a carnival ride that moves in a circle. In this activity, you will determine how an object on a curved path moves.

Make a Prediction How will a curved track affect the direction a marble travels? What will happen when a marble is rolled on an open floor?

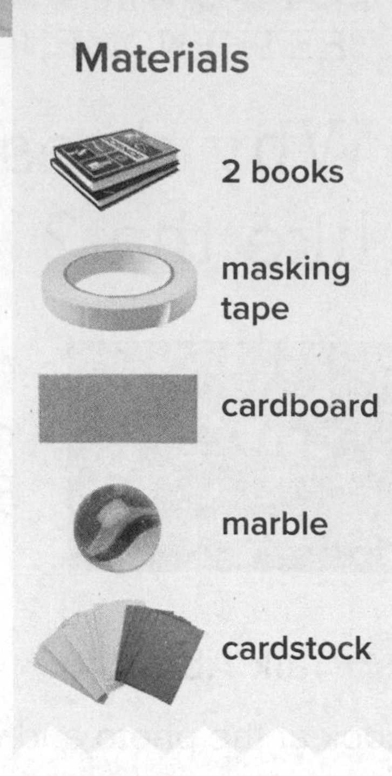

Carry Out an Investigation

1. Working with your group, use the materials to create a ramp that will cause the marble to move in a curved path.
2. Place the marble at the top of the ramp. Release the marble and observe the direction it moves.
3. **Record Data** In your data table, record the direction the marble moved on the curved ramp.
4. Now clear a space on the floor, and place the marble on the floor.
5. Revisit your prediction about the direction the marble will move when you roll it on the open floor.
6. **Record Data** Push the marble, and record the results in the data table.

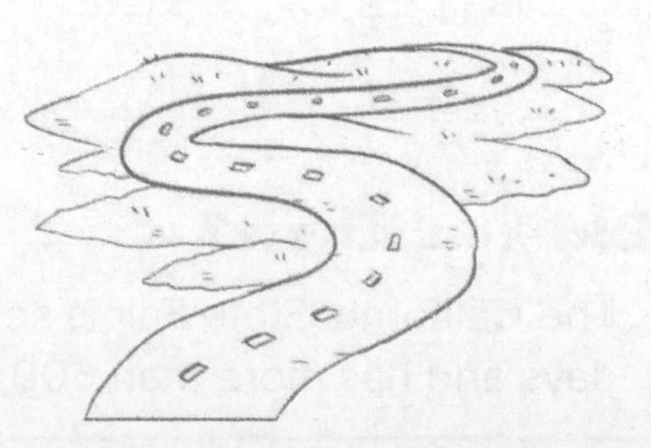

	observations
curved ramp	
open floor	

Communicate Information

7. Compare the way the marble moved on the ramp and on the open floor.

8. Explain why it was hard to predict the direction the marble would go on the open floor.

Talk About It

What would happen if you created a ramp that started in the same place and split into two different ramps? Which way would the marble roll?

VOCABULARY

Look for these words as you read:

direction

distance

motion

position

speed

Describe an Object's Position

Position Think back to the *Moving Marble* activity. What position was the marble in when you started? Where did the marble end up after it was pushed down the ramp? When you describe the position of something, you compare it to the objects around it. **Position** is the location of an object. You can use words like *above, below, next to,* and *far away from* to describe the position of an object. Look at the boy on the beach. Draw an X on the inner tubes that are above the polka-dotted pink inner tube.

Distance The amount of space between two objects or places is **distance**. Millimeters, centimeters, meters, and kilometers are examples of units used in the metric system to measure distance. In the US customary system, distance might be measured in inches, yards, or miles. You can use a ruler or a meterstick to measure distance.

Direction When describing position, you must use both distance and direction. **Direction** tells which way a line points from one object or place to another. The words *north, south, east,* and *west* describe direction. You can also use words such as *left, right, up, down, forward,* and *backward* to describe direction. With a partner, describe the position of the pink inner tube on the previous page, using distance and direction.

1. **MATH Connection** How could you find the distance from your desk to the door?

2. Draw a diagram of the *Moving Marble* activity. Label the position of the marble at the start and end. Draw an arrow to identify the direction the marble rolled. Label the distance from start to stop with a line.

Motion

> **GO ONLINE** Watch the video *Patterns in Motion* to compare how different things move.

Look at the pictures of the dog in different positions. First, you can see that the dog is on the ground. Next, you see the dog come completely off the ground. What happened to the dog? It moved. You know that the dog moved because its position changed. **Motion** is the process of changing position.

You can observe motion in different ways. Some objects move in a straight line. Other objects can move round and round, back and forth, or in a zigzag pattern.

> **GO ONLINE** Explore *Draw the Pattern* to learn more about different patterns.

Measuring Motion

Distance There are many ways that you can measure motion. One way is to measure the distance that an object moves. As you learned on the previous page, distance is the measurement between an object's starting position and its current position. You can measure larger distances in units such as meters, yards, miles, or kilometers.

Time Suppose it took you three minutes to walk from your classroom to the playground yesterday. Today it took you five minutes to walk to the playground. You moved the same distance, but your motion today took more time. The time it takes to move a distance is one way to describe motion.

Speed Distance and time can be used to find speed. **Speed** is the measure of how fast or slow something moves. An object that is moving fast goes a distance in a short amount of time. It takes a longer time for a slower object to move the same distance.

REVISIT Revisit the Page Keeley Science Probe on page 5.

PAGE KEELEY SCIENCE PROBES

Direction Direction points out the path from one position to another. Suppose you walk the length of a soccer field. Then, you turn around and walk the whole length back. Is your motion the same both times? Even if the distance and time are the same, the motion is different. You walk different directions back and forth across the field. The direction an object moves is part of the way you describe its motion.

Predicting Motion

Measurements of motion may help you predict future motion. Look at the picture of the girl on the swing. You can predict when she will change direction. You can also predict how much time it will take her to swing back and forth. Draw an arrow predicting the direction the girl will swing next.

What is the **pattern of motion** when you are on a swing? How can you **predict the patterns of motion** of the girl swinging in the photo?

Reread the first and second paragraph on page 12. Circle the different types of motion. Label the different types of motion using words and arrows to show the direction.

GO ONLINE Explore the PhET simulation *Forces and Motion Basics*. Collaborate with a partner. What patterns do you see?

What Does a Statistician Do?

Statisticians collect and study data to help solve real-world problems. Statisticians are hired to study data in business, education, and other fields. With the right data, there is often enough information for people to find solutions to their problems.

Statisticians who work for car companies might collect data about the speed of the cars–how quickly they accelerate, how far they can go, and how long it takes. Statisticians who work for universities might collect data about how long it takes students to graduate.

Talk About It

How do you think a statistician might be involved in designing a skatepark?

It's Your Turn

Think like a statistician. Complete the activity on the next page and explore the relationship between distance and time.

INQUIRY ACTIVITY

Hands On

Movement of a Wind-Up Toy

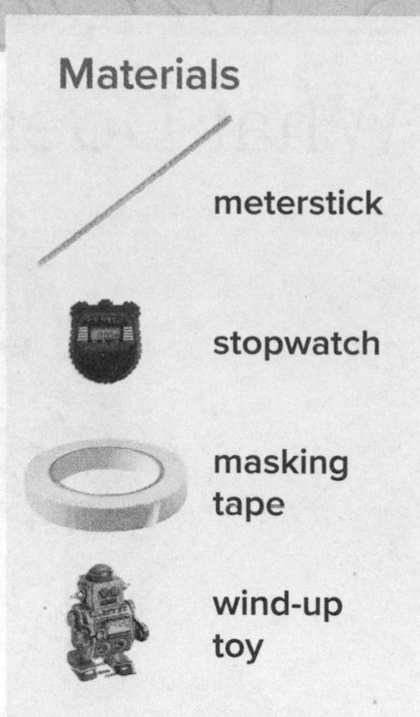

You have learned that you can measure the space between an object's starting position and its new position. Now let's look at the distance an object, such as a wind-up toy, travels in a given amount of time.

Make a Prediction How far will the wind-up toy travel in 10 seconds? How far will it travel in 20 seconds? Record your prediction in centimeters.

____

____

____

Carry Out an Investigation

1. Place a piece of tape on your desk. This is the starting line.
2. Wind up the toy and place it at the starting line. Be sure to wind up the toy all the way. Get your stopwatch ready.
3. **Record Data** As you release the toy, start the timer. Stop the toy after 10 seconds, and mark the spot with tape. Use a meterstick to measure how far the toy went.
4. Repeat two more times, and write the distance the toy traveled for each trial.
5. Repeat this activity, but this time, stop the toy after 20 seconds.

	Trial 1	Trial 2	Trial 3
Distance Traveled in 10 seconds			
Distance Traveled in 20 seconds			

Communicate Information

6. Do your findings support your prediction? Explain.

7. Look at the data you collected. Compare the distance the toy traveled in 10 seconds and 20 seconds. What patterns do you notice?

8. How far do you think the wind-up toy will travel in 25 seconds? Explain your reasoning.

Talk About It

You used a meterstick and stopwatch to measure distance and time during the activity. Discuss other tools and strategies you can use to measure the motion of the toy.

LESSON 1

Review

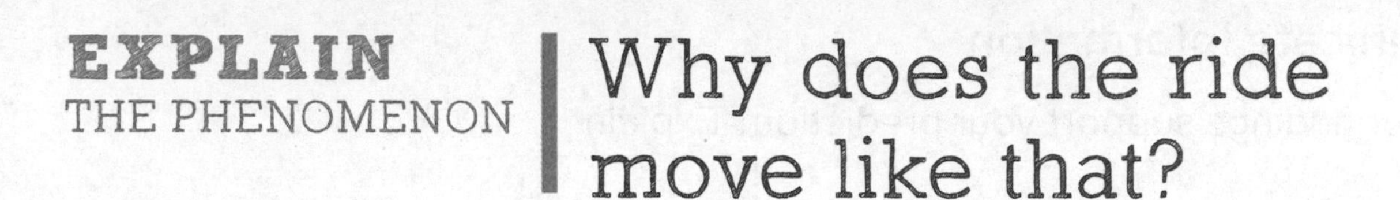

Summarize It

Explain patterns in motion. How can you measure the patterns of motion? What patterns in motion can you observe and measure in your classroom?

Revisit the Page Keeley Science Probe on page 5. Has your thinking changed? If so, explain how it has changed.

Three-Dimensional Thinking

1. A ball is moving in a zigzag pattern, but you need it to go straight to reach the goal. How do you change the ball's motion?

A. I leave it alone. It will go straight when I want it to.

B. I have to push the ball in a straight line to make it reach the goal.

C. I have to pull the ball toward me and bounce it on the ground toward the goal.

2. The data below shows the distance a toy car traveled down three different ramps.

	Ramp 1	Ramp 2	Ramp 3
Distance Traveled in 20 seconds	**4 cm**	**12 cm**	**5 cm**

Which ramp is most likely the tallest? Explain how you know.

____

____

____

3. Two objects start at the same location and travel at the same speed for one minute, but they end up in different locations. How did their motions differ?

____

____

____

____

____

Extend It

You are a principal of a school and need to have the gym set up for a magic show. Use direction, position, and distance to describe your gym setup. Once completed, give your plan to a classmate to draw your setup on a separate sheet of paper. Explain the outcome below.

OPEN INQUIRY

What questions do you still have?

Plan and carry out an investigation to answer one of your questions.

KEEP PLANNING

STEM Module Project
Engineering Challenge

Now that you have learned about position and motion, go to your Module Project to explain how the information will help you build your skatepark.

LESSON 2 LAUNCH

Golf Ball

Three friends are playing golf. They each have different ideas about the forces that act on a golf ball. This is what they think:

Finn: *Forces act on the golf ball only when the golfer hits the ball.*

Pete: *Forces act on the golf ball only when the ball is on the tee.*

Tad: *Forces act on the golf ball when it is on the tee and when the golfer hits the ball.*

Who has the best idea about forces? ______________________

Explain why you think it is the best idea.

____

____

____

____

You will revisit the Page Keeley Science Probe later in the lesson.

LESSON 2

Forces Can Change Motion

Vstock/Alamy Stock Photo

ENCOUNTER
THE PHENOMENON

How are they going down the slide so fast?

GO ONLINE

Check out *Slides* to see the phenomenon in action.

Talk About It

Look at the photo and watch the video of the kids going down the slide. What questions do you have about the phenomenon? Talk about your questions and observations with a partner.

Did You Know?

London has the longest and tallest slide in the world. It takes about 40 seconds to go down!

INQUIRY ACTIVITY

Hands On

Forces Affect the Way Objects Move

You saw people going down a slide. A slide is one kind of ramp. Investigate how the height of a ramp will change a toy car's motion.

Make a Prediction How will the height of a ramp affect the motion of a toy car?

Carry Out an Investigation

1. Stack two books on the floor. Lean a piece of cardboard along the top book to make a ramp. Tape the edge of the cardboard to the floor.
2. Place a toy car at the top of the ramp. Release the car.
3. MATH Connection Use the meterstick to measure the distance the car traveled.
4. **Record Data** Record the distance the car traveled in the data table.
5. Repeat steps 2–4 for a total of three trials.

6. Repeat steps 1–5 with a stack of four books.

	Distance Traveled in Centimeters		
	Trial 1	**Trial 2**	**Trial 3**
Two-book ramp			
Four-book ramp			

7. Compare the distances the toy car traveled with the two ramps. What pattern do you see?

8. Predict what would happen if your ramp had six books.

INQUIRY ACTIVITY

Communicate Information

9. Did your observations support your prediction? Explain.

10. Draw a real-world example of how the height of a ramp affects the motion of an object.

MAKE YOUR CLAIM

What makes a toy car slide down a ramp?

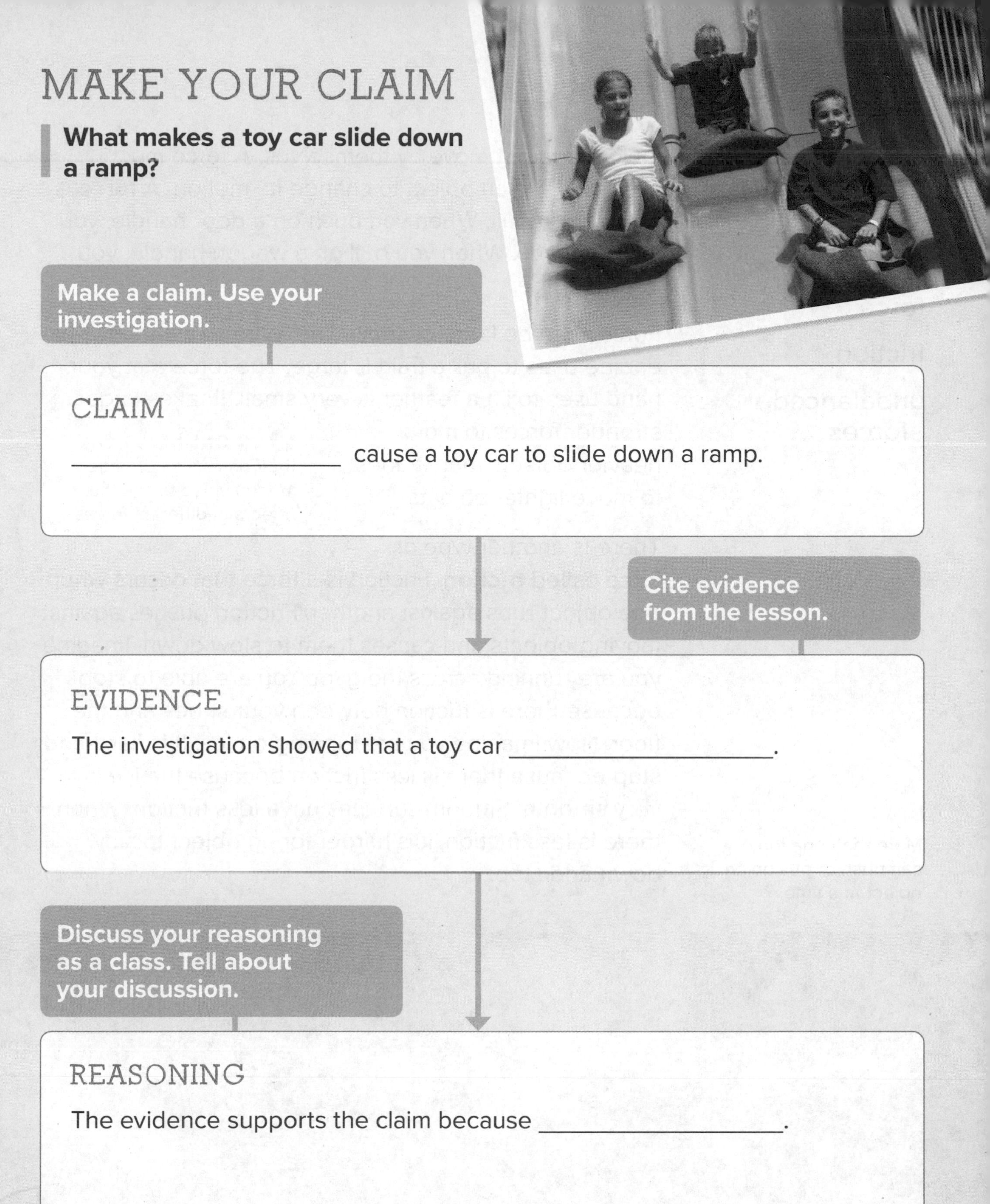

Make a claim. Use your investigation.

CLAIM

____________________ cause a toy car to slide down a ramp.

Cite evidence from the lesson.

EVIDENCE

The investigation showed that a toy car ____________________.

Discuss your reasoning as a class. Tell about your discussion.

REASONING

The evidence supports the claim because ____________________.

You will revisit your claim to add more evidence later in this lesson.

VOCABULARY

Look for these words as you read:

balanced forces

force

friction

unbalanced forces

Forces

Objects do not move by themselves. A force must be applied to an object to change its motion. A **force** is a push or a pull. When you push on a door handle, you apply a force. When you pull on a wagon handle, you apply a force.

Forces can be large or small. The force that a train engine uses to pull a train is large. The force that your hand uses to lift a feather is very small. It takes larger, stronger forces to move heavier objects than it does to move lighter objects.

GO ONLINE Watch the *Forces Can Change Motion* video to see the effects of different forces.

There is another type of force called **friction**. Friction is a force that occurs when one object rubs against another. Friction pushes against moving objects and causes them to slow down. Imagine you are running across the gym. You are able to stop because there is friction between your shoes and the floor. Now imagine you are running on ice. It is harder to stop because there is less friction because the ice is very smooth. Smooth surfaces have less friction. When there is less friction, it is harder for an object to slow down and stop.

More than one force can push or pull on an object at a time.

push

pull

Balanced and Unbalanced Forces

Forces can set objects into motion. When you put a heavy backpack on your desk, the backpack does not move. Gravity pulls the backpack toward Earth, but your desk pushes up on the backpack with a force. The strength of that force is exactly equal to the pull of gravity. The forces are balanced.

Balanced forces are forces that cancel each other out when acting together on an object. Sometimes balanced forces are equal in size and opposite in direction, but forces do not have to be equal and opposite to be balanced. When an object is sitting still, all the forces acting on it are balanced. However, when objects are moving at a constant velocity, or speed, they are also balanced. Balanced forces do not cause a *change* in motion.

Suppose you push that heavy backpack across your desk. The backpack is moving. This is due to **unbalanced forces**. Forces that are not equal to each other are called unbalanced forces. If there is more than one force acting on an object, then the total of all the forces determines the direction of motion.

The forces applied to the stuffed bear are balanced, it is not moving.

The dogs are applying a greater force to the sled, so the sled is moving.

Talk About It

Which forces cause a change in motion?

Changing Motion

GO ONLINE Explore the PhET simulation *Forces and Motion* to see the forces in tug of war.

Think back to your toy car. With a partner, brainstorm five ways you can make an object have motion. In the table, draw a picture using arrows to indicate direction. Label what force was applied and if the forces were balanced or unbalanced.

Motion	Forces Acting on Object	Balanced or Unbalanced
Make an object remain still		☐ **Balanced** ☐ **Unbalanced**
Make an object move forward		☐ **Balanced** ☐ **Unbalanced**
Make an object move faster, forward		☐ **Balanced** ☐ **Unbalanced**

Motion	Forces Acting on Object	Balanced or Unbalanced
Make an object move sideways		☐ Balanced ☐ Unbalanced
Make an object move up		☐ Balanced ☐ Unbalanced
Make an object move down		☐ Balanced ☐ Unbalanced

REVISIT PAGE KEELEY SCIENCE PROBES

Revisit the Page Keeley Science Probe on page 21.

Inspect

Read the passage *Skateboarding*. Underline text evidence that tells what two things a skateboarder needs to do tricks.

Find Evidence

Reread How does a skateboarder get high enough to do a trick? Highlight the text that explains.

Notes

Skateboarding

Skateboarding is a sport that began in 1950 in California. Before there were skateparks, skateboarders practiced in empty swimming pools. Today, there are hundreds of thousands of skateparks in the United States.

Skateboarding is a fun sport that requires only a few pieces of equipment. A skateboard and protective gear makes someone ready to hit the park. Although skateboards can vary and have unique designs, all are made of three basic parts: a board, wheels, and trucks, which connect the wheels to the board and allow the board to turn.

To be safe, skateboarders have to wear helmets to protect their heads. They also wear gear to protect their wrists and knees.

There is science involved in designing a skatepark. Architectural designers apply principles of motion and force so that the skateboarders can get the speed they need.

Notice the many slopes of the skatepark in the photo on this page. When skateboarders push themselves down a slope, their speed increases. They go across a flat surface as they stand on the board. Leaning their body to one side or the other causes the wheels to move the direction they want, right toward another slope. Their speed remains the same because balanced forces are acting on the skateboard. When they go down another slope, they use unbalanced forces to increase their speed to carry them across another flat surface and up the next slope. With enough speed, they can get high enough to do their tricks in the air.

Make Connections

Talk About It

What does an architectural designer need to know about skateboarders in order to design a skatepark?

Notes

FOLDABLES®

Cut out the Notebook Foldables tabs given to you by your teacher. Glue the anchor tabs as shown below. Use what you have learned throughout the lesson to describe the picture using vocabulary words.

Glue anchor tab here

STEM Connection

What Does a Landscape Designer Do?

Landscape Designers plan and design public spaces, residential areas, and college campuses. They are creative people who like to work on big projects. You might think landscape designers work only with plants and lawns, but they also know a lot about paving, walls, fencing, wood, concrete, and metal. They know about irrigation and water management, too.

Landscape designers also think a lot about motion and force. When they design spaces where people will work or play, they consider what objects will move through the spaces and the forces that will affect the movement of the objects.

It's Your Turn

As a landscape designer, what information would you need to build a skatepark? How could you find out how skateboarders move in a park, and how would your findings influence your design?

INQUIRY ACTIVITY

Hands On

On the Move

When playing with toy cars, some cars are faster than others. With a push on the floor, the car starts out fast. It then slows down and stops. Investigate how different materials can affect the speed and distance of a toy car.

Make a Prediction What would happen if a toy car rolls over different materials?

____

____

____

Materials

- 4 books
- cardboard
- masking tape
- toy car
- meterstick
- sandpaper
- cotton cloth

Carry Out an Investigation

1. Make the four-book ramp. Copy the data from the "Four-book ramp" row of the table on page 25 into the "Floor" row of the table on page 37.
2. Tape a layer of sandpaper at the bottom of the cardboard ramp. Release the car from the top of the ramp.
3. **Record Data** Measure and record the distance the car travels. Repeat for a total of three trials.
4. Remove the sandpaper. Tape a cotton cloth to the floor at the bottom of the cardboard ramp. Release the car from the top of the ramp.
5. **Record Data** Measure and record the distance the car travels. Repeat for a total of three trials.

	Distance Traveled		
	Trial 1	**Trial 2**	**Trial 3**
Floor			
Sandpaper			
Cotton cloth			

Communicate Information

6. Why did the car slow down when traveling on a sandpaper surface?

Talk About It

Compare your results with your classmates' results. What material would you use if you wanted an object to stop quickly? Why do you think some materials caused more friction than other materials?

COLLECT EVIDENCE

Add evidence to your claim on page 27 about how forces affect an object's motion.

EXPLAIN THE PHENOMENON

How are they going down the slide so fast?

Summarize It

Explain the effects of a force acting on an unmoving object.

Revisit the Page Keeley Science Probe on page 21. Has your thinking changed? If so, explain how it has changed.

Three-Dimensional Thinking

1. How do forces change the motion of objects?

A. Forces can change the speed or direction of an object's motion.

B. The size of the force affects the speed of the object.

C. The direction of the force affects the direction of the object's motion.

D. All the above

E. None of the above

2. An egg is about to roll off the counter. How can you get the egg to stop without picking it up?

3. Explain why the amount of friction would be different on an icy surface and a dry, concrete surface. How does the amount of friction affect the movement of an object across both surfaces?

Extend It

You are the mayor of San Francisco, California. The trolley cars are in need of repair. How might you communicate with your citizens about the importance of repairing the cable car brakes? Think about what you have learned in this module to help explain force and motion.

Write a speech, draw a poster, create a flyer, or use media.

____

____

____

____

____

____

____

____

____

____

____

KEEP PLANNING

STEM Module Project
Engineering Challenge

Now that you have learned how forces can affect motion, go to your Module Project to explain how the information will affect your plan for the skatepark.

STEM Module Project
Engineering Challenge

Design a Skatepark

You have been hired as an architectural designer. Using what you have learned throughout this module, you will design a skatepark. Your goal will be to design, build, and test a model that will successfully get a marble from one end of the park to the other using parameters set by your teacher.

Planning after Lesson 1

Apply what you have learned about motion to your project planning.

How does knowing about motion affect your project planning?

Record information to help you plan your model after each lesson.

STEM Module Project
Engineering Challenge

Planning after Lesson 2

Apply what you have learned about forces that can change motion.

What factors should be considered when building your model of a skatepark?

Research the Problem

Research building designs by reading the Investigator article *Play It Safe!* Go online to teacher-approved websites, or by finding books on designing skateparks at your local library.

Source	Information to Use in My Project

Sketch Your Model

Draw your ideas on a separate piece of paper. Select the best one to build and test.

Design a Skatepark

Look back at the planning you did after each lesson.
Use that information to complete your final module project.

The Engineering Design Process

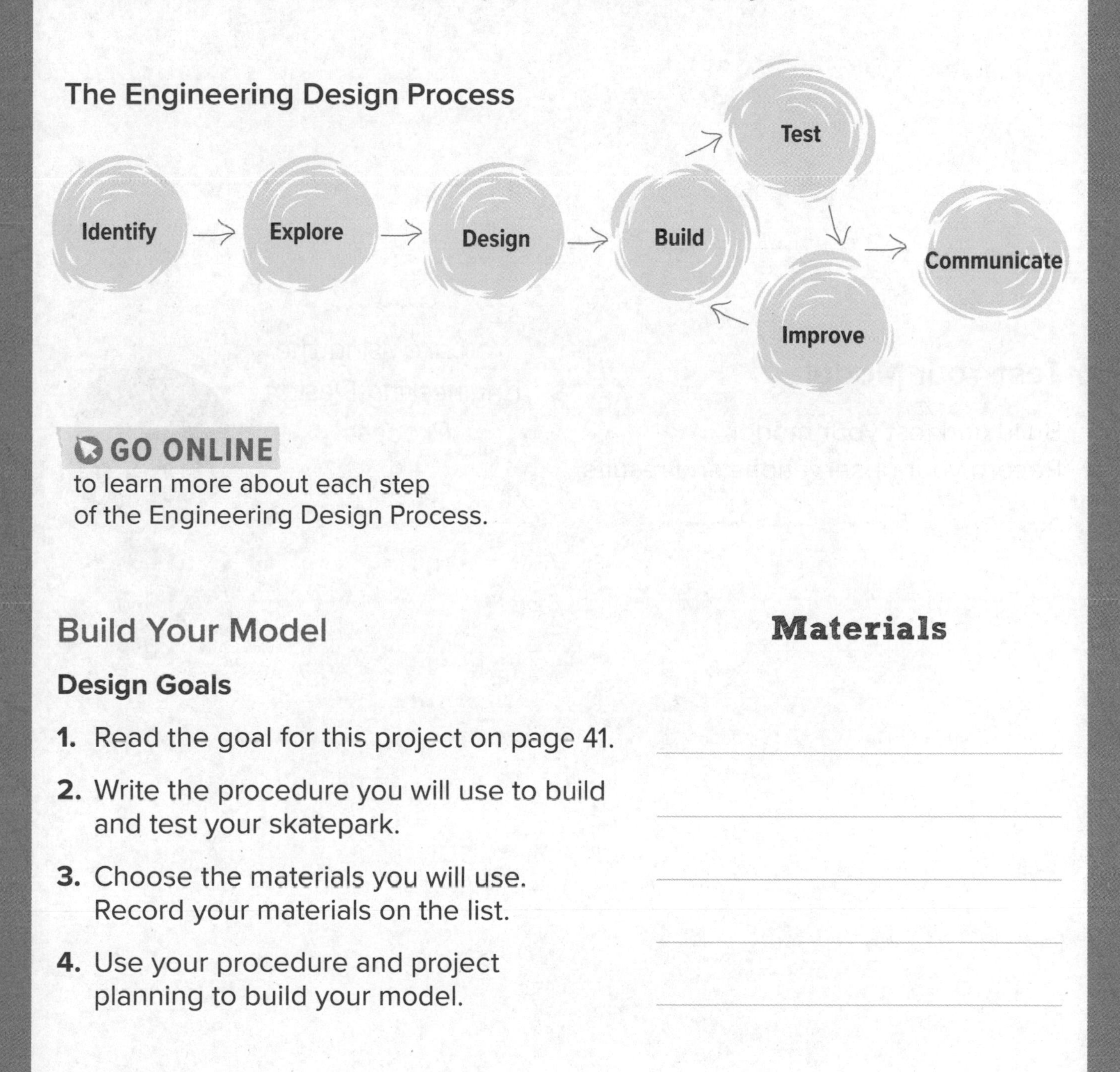

GO ONLINE to learn more about each step of the Engineering Design Process.

Build Your Model

Design Goals

1. Read the goal for this project on page 41.
2. Write the procedure you will use to build and test your skatepark.
3. Choose the materials you will use. Record your materials on the list.
4. Use your procedure and project planning to build your model.

Materials

Procedure:

Test Your Model

Build and test your model.
Record your observations and results.

MODULE WRAP-UP

REVISIT THE PHENOMENON

Describe how the results of your project can help you explain how the skateboarder is moving.

Revisit your project if you need to gather more evidence.

How has your thinking changed? Explain.

Electricity  and Magnetism

ENCOUNTER
THE PHENOMENON

What is moving those objects?

Electromagnets

GO ONLINE

Check out *Electromagnets* to see the phenomenon in action.

Talk About It

Look at the photo and watch the video of the objects being moved. What questions do you have about the phenomenon? Talk about your observations with a partner.

Did You Know?

One of the world's most powerful magnets is at Los Alamos National Laboratory in New Mexico.

STEM Module Project Launch
Engineering Challenge

Design a Self-Closing Gate

Lesson 1
Electricity and Designing Solutions

The gate on the fence of the school garden is often left open. How can you use electricity and magnetism to make a self-closing gate? You are being hired as an engineer to design and build a self-closing gate. By the end of this module, you will identify the criteria and materials needed to make a self-closing gate.

Engineers design new products that solve problems. Sometimes they improve designs that are not working properly.

Lesson 2
Magnetism and Designing Solutions

ANTONIO
Robotics Engineer

STEM Module Project

Plan and Complete the Engineering Challenge Use what you learn throughout the module to complete the challenge.

LESSON 1 LAUNCH

Salma's Hair

Salma rubbed a balloon on her hair. She then held the balloon over her head. Salma and her friends laughed when her hair went straight up and stuck to the balloon. They each had different ideas about why Salma's hair went up toward the balloon. Here is what they said:

Salma: *I think the balloon is acting like a magnet on my hair.*

Curt: *I think the balloon and hair are electrically charged.*

Nicki: *I think the balloon has tape that pulled the hair.*

Who do you agree with most? ______________________

Explain why you agree.

____

____

____

____

____

____

You will revisit the Page Keeley Science Probe later in the lesson.

LESSON 1

Electricity and Designing Solutions

ENCOUNTER
THE PHENOMENON

What does electricity have to do with a balloon attracting hair?

GO ONLINE

Check out *Static Electricity* to see the phenomenon in action.

Talk About It

Look at the photo and watch your teacher demonstrate with a balloon. What questions do you have about the phenomenon? Talk about your observations with a partner.

Did You Know?

Just like the balloon, lightning occurs because of buildup of electrical charge in the clouds!

INQUIRY ACTIVITY

Hands On

Static Charge

In the Encounter the Phenomenon video, you observed hair rising and falling as a result of static electricity. Think about how the balloon affects other materials.

Make a Prediction What will happen to a balloon, paper confetti, running water, and gelatin when a balloon that has been rubbed with wool comes near them?

Carry Out an Investigation

1. Rub one inflated balloon on a piece of wool.
2. Hold the balloon close to the other balloon.
3. **Record Data** Record your observations.
4. Rub a balloon on a piece of wool and slowly bring the balloon close to the paper confetti. Record your observations.
5. Rub a balloon. Hold the balloon above a bowl of gelatin powder. Record your observations.
6. Rub a balloon. Hold the balloon near a thin, steady stream of water. Record your observations.
7. Rubbing the balloon each time, try to stick the balloon to three different surfaces in your classroom. Record the surfaces and your observations.

Materials

Material	Observations
Balloon	
Confetti	
Gelatin	
Water	

Communicate Information

8. Did the results of your investigation support your prediction? Explain.

9. What other questions do you have about this activity?

OPEN INQUIRY

10. Choose one of your questions and write a procedure you can use to find the answer.

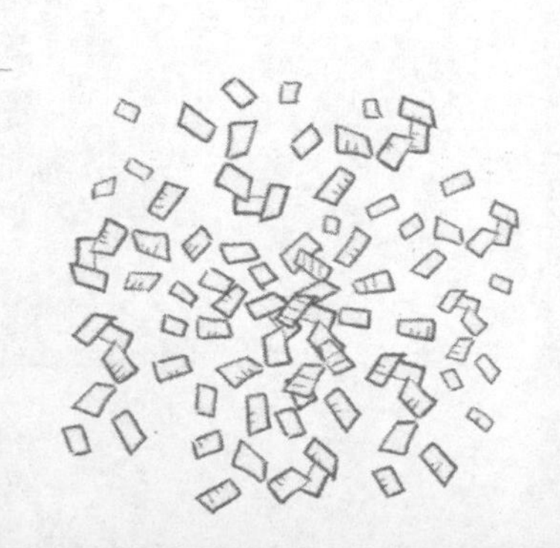

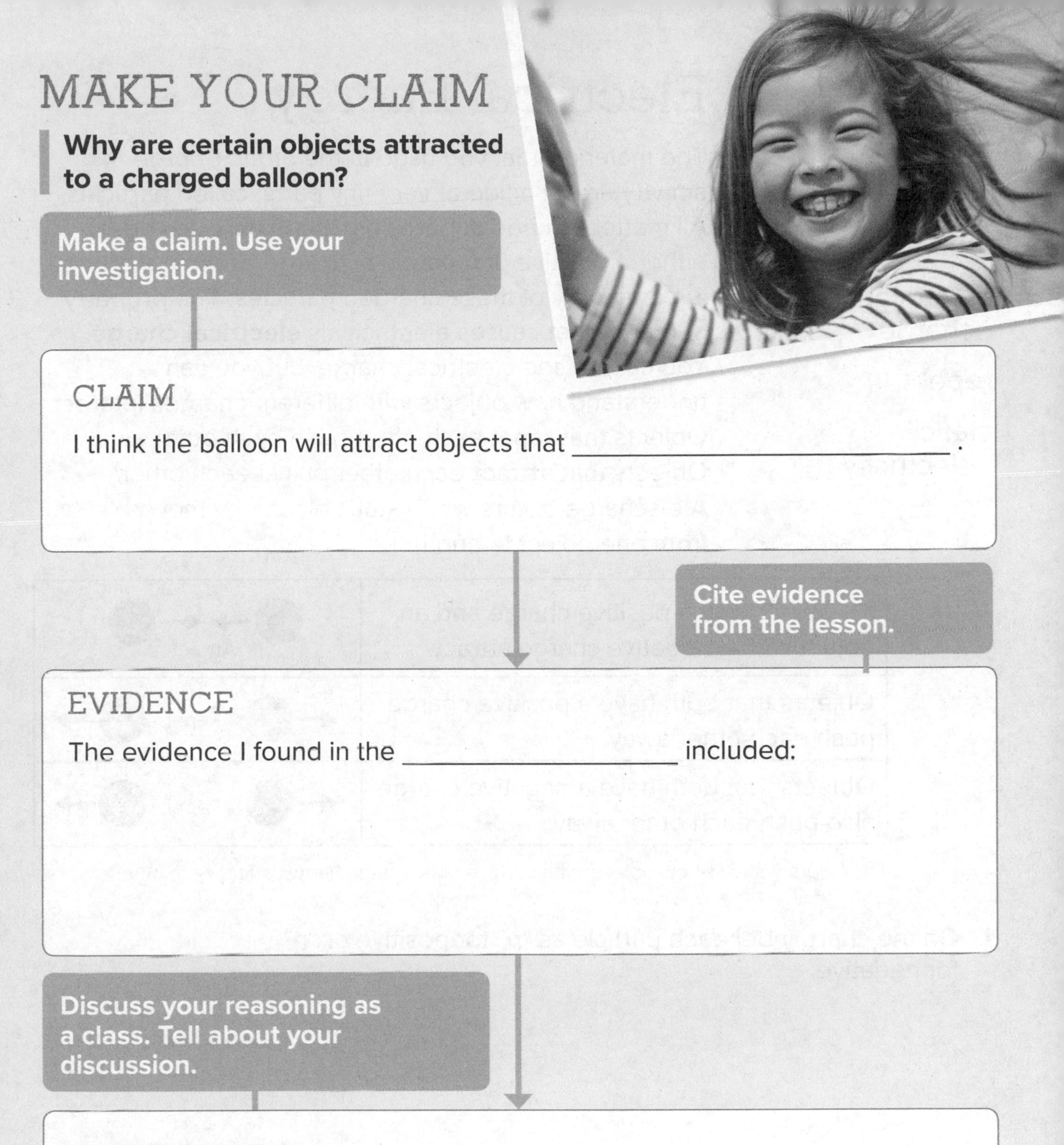

MAKE YOUR CLAIM

Why are certain objects attracted to a charged balloon?

Make a claim. Use your investigation.

CLAIM

I think the balloon will attract objects that ______________________.

Cite evidence from the lesson.

EVIDENCE

The evidence I found in the ________________ included:

Discuss your reasoning as a class. Tell about your discussion.

REASONING

My reasoning for my claim is ________________.

You will revisit your claim to add more evidence later in this lesson.

VOCABULARY

Look for these words as you read:

attract

electrical charge

repel

static electricity

Electrical Energy

The materials that you used in the *Static Charge* activity are all made of very tiny parts, called particles. All matter is made of particles. Some particles have either a positive or a negative charge. Electrical energy is the energy of these charged particles. The property of matter that causes electricity is **electrical charge**. You cannot see electrical charge, but you can understand how objects with different charges interact. Objects that **repel** each other push each other away. Objects that **attract** each other pull at each other. A discharge occurs when static electricity moves from one object to another.

An object with a positive charge and an object with a negative charge attract.	Attract
Objects that both have a positive charge push each other away.	Repel
Objects that both have a negative charge also push each other away.	Repel

The diagram shows how objects with positive or negative charges affect each other.

1. On the chart, label each particle as "p" for positive or "n" for negative.

2. Explore the simulation. What happens when you rub the balloon on the sweater?

GO ONLINE Explore the PhET simulation *Balloons and Static Electricity*.

Static Electricity

All objects are made of charged particles. Most objects have the same number of positive particles and negative particles. When they do, the charges are balanced. When two objects touch, negative particles can move from one object to the other. Negative particles may build up on one object. That object has a negative charge. A buildup of electrical charge is called **static electricity**.

Think back to the *Static Charge* activity. After the balloon was rubbed, it had more negative particles. Those negative particles were then attracted to the positive particles in some of the objects and were repelled if the object also had a buildup of negative particles.

If you hold a charged balloon near a wall, the negative charge attracts the positive (+) particles on the wall. This attraction causes the balloon to stick to the wall.

Think back to the *Static Charge* activity. Why did you have to rub the balloon between each object?

REVISIT

Revisit the Science Page Keeley Probe on page 49.

Inspect

Read the passage *Electric Current.* Underline text evidence that tells how people use electric currents.

Find Evidence

Reread Highlight text that tells what a circuit needs to have energy flow.

Notes

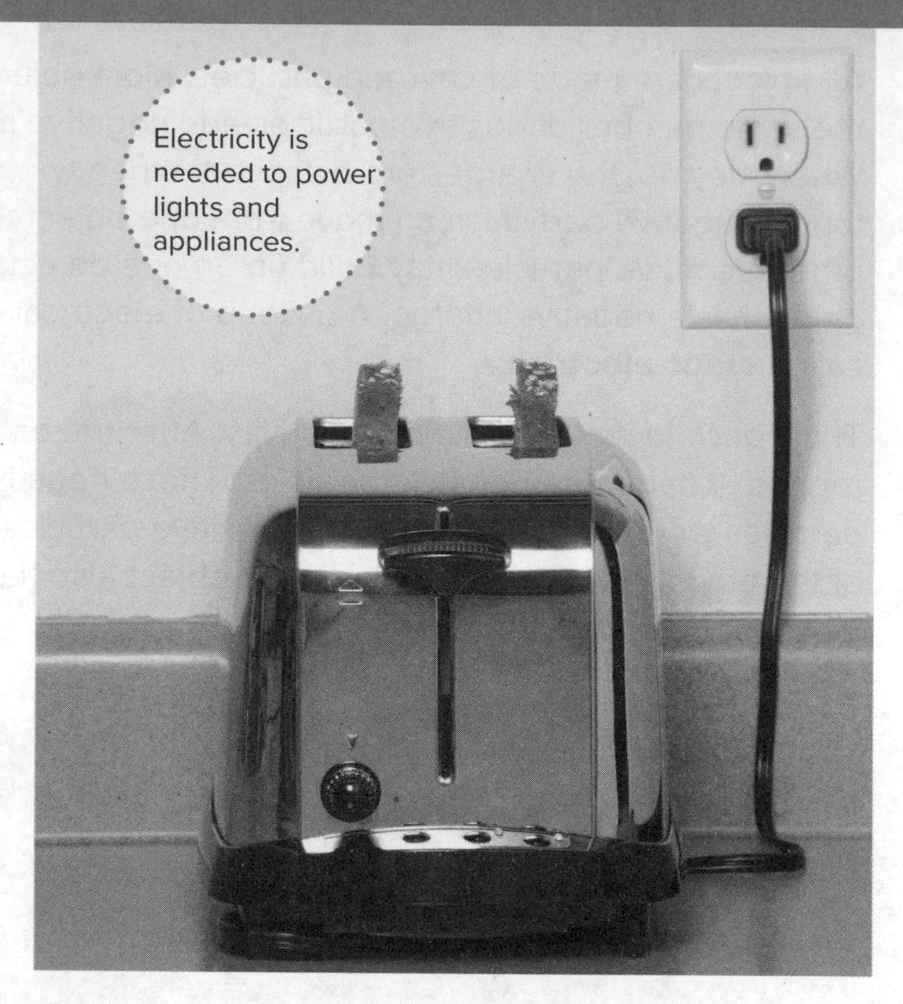

Electric Current

Charge particles can build up on an object. They can also be made to flow. A flow of charged particles is called an electric current. You use electric current every day. Electric current provides the energy you need to power lights, radios, computers, and many other products. We use energy from electric current to produce light, sound, and motion.

This is an example of a simple circuit using a battery.

Make Connections

Talk About It

Think about how energy flows in static electricity. Compare it to how energy flows in a circuit.

Notes

Electric current needs a path through which to flow. A circuit is a path that is made of parts that work together to allow current to flow. Simple circuits have several parts. A battery or an electric outlet may be the circuit's source of power. Wires connect the different parts of the circuit. These wires are usually made of copper or another type of metal, and are wrapped in plastic. The last part needed to complete a simple circuit is a load. A load is the device that needs an electric current to work.

People use the energy flowing through electrical currents every day. Look around your classroom. What do you see that uses electricity?

Thomas Edison

Thomas Edison has been described as America's greatest inventor. He improved the design of the electric lightbulb. He used the designs of other scientists to make the lightbulb cheaper, last longer, and use less electric current.

Edison also designed a way to provide electricity to street lamps and nearby houses.

PRIMARY SOURCE

Why do you think making a light bulb that was cheaper, lasted longer, and used less electric current was so important?

____

____

____

COLLECT EVIDENCE

Add evidence to your claim on page 55 about why certain objects are attracted to balloons.

FOLDABLES®

Cut out the Notebook Foldables given to you by your teacher. Glue the anchor tabs as shown below. Describe the picture using vocabulary words.

STEM Connection

What Does an Electrical Engineer Do?

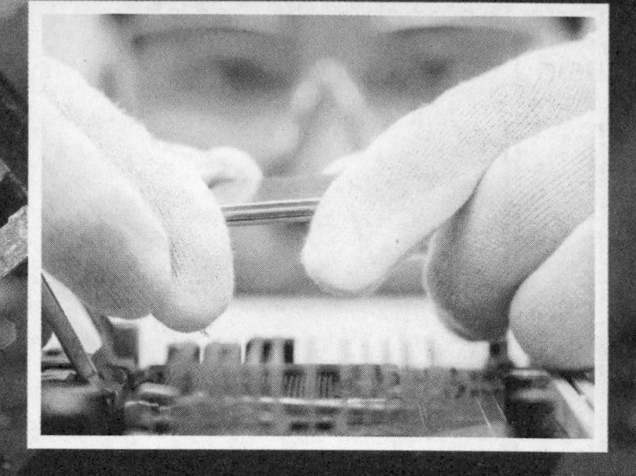

Electrical Engineers design electrical devices like cell phones, computers, and robots. They also design things like GPS technologies. Electrical engineers are creative people who like to work with other people to figure out how electricity can be helpful.

Electrical engineers work with tiny microchips, and they also work with large power station generators. When working on computers, the room must be "clean". This room must be free of dust, hair, lint, and static electricity to protect the equipment.

Have you ever had static electricity in your shirt and it stuck to you? Static electricity happens when an electrical charge builds up on the surface of an object. We call it *static* because the electrical charge stays in one place.

It's Your Turn

How does electricity affect an object's motion? As an electrical engineer, how could you use static electricity to design a way to keep a playground gate closed?

INQUIRY ACTIVITY

Hands On

Eliminate Static Electricity

Have you ever found a sock stuck to your shirt? The sock is stuck due to static electricity. Some things can remove static electricity.

Make a Prediction Will water, a dryer sheet, or a polyester cloth remove static electricity best?

____

____

Materials

3 balloons

wool cloth

paper confetti

water in spray bottle

dryer sheet

polyester cloth

Carry Out an Investigation

1. Rub an inflated balloon with a wool cloth.
2. **Record Data** Hold the balloon just above a pile of confetti. Record your observations on a separate piece of paper.
3. Spray the balloon with water. Then, hold it over the pile of confetti. Record your observations.
4. Using a new balloon, repeat steps 1–3 using a dryer sheet instead of water.
5. Using a new balloon, repeat steps 1–3 using a polyester cloth instead of water.

Communicate Information

6. Did the results of your investigation support your prediction? Explain.

____

Talk About It

What property of the material made it eliminate the static electricity?

LESSON 1

Review

EXPLAIN THE PHENOMENON

What does electricity have to do with a balloon attracting hair?

Summarize It

Explain how electricity is causing the girl's hair to stand up.

REVISIT PAGE KEELEY SCIENCE PROBES

Revisit the Page Keeley Science Probe on page 49. Has your thinking changed? If so, explain how it has changed.

Three-Dimensional Thinking

1. What is an electrical charge?

A. the property of matter that causes electricity

B. the uninterrupted flow of electricity

C. a sudden burst of energy

D. the path that allows electrical current to flow

2. If you rub two balloons with a wool cloth,

A. the balloons will attract each other.

B. the ballons will not affect each other.

C. the balloons will repel each other.

D. the balloons will pop.

3. Objects with the same charge ________________ each other.

A. attract

B. balance

C. circuit

D. repel

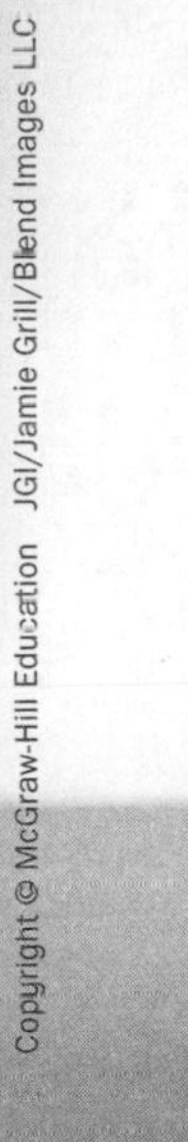

Extend It

OPEN INQUIRY

Recall how a small amount of static electricity produced by a balloon was able to attract small objects.

How can you use a larger amount of static to solve a problem? Come up with a plan below. First state the problem. Then, design a device or a plan that uses static electricity to solve the problem.

KEEP PLANNING

STEM Module Project

Engineering Challenge

Now that you have learned how electricity can move objects, go to your Module Project to explain how the information will affect your plan for your gate.

LESSON 2 LAUNCH

Magnet and Paper Clip

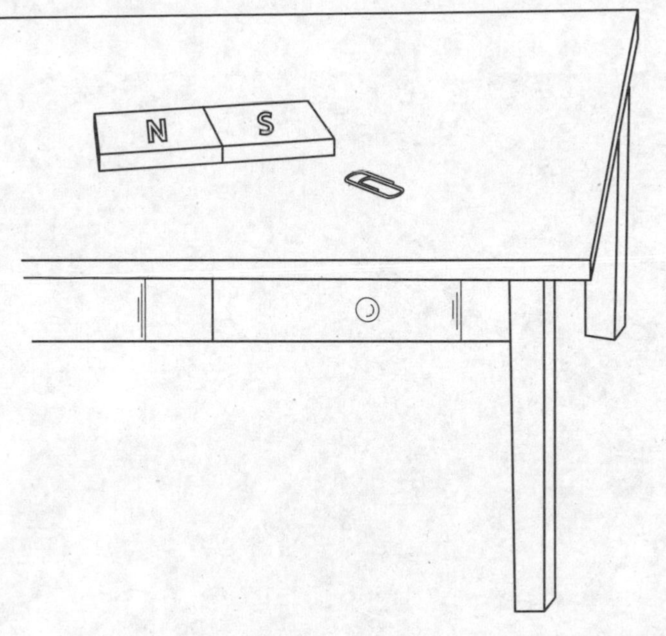

Magnets are used to attract objects. They move toward the magnet. Magnets are also used to repel objects. They move away from the magnet. What happens when a bar magnet is placed near a steel paper clip? Circle the best answer.

A. *Both ends of the magnet will attract the paper clip.*

B. *One end of the magnet attracts the paper clip. The other end repels the paper clip.*

C. *Both ends of the magnet repel the paper clip.*

Explain your thinking. How did you decide what happens with both ends of the magnet?

You will revisit the Page Keeley Science Probe later in the lesson.

LESSON 2

Magnetism and Designing Solutions

ENCOUNTER
THE PHENOMENON

How is that object being moved without anything touching it?

GO ONLINE

Check out *Magnetic Interaction* to see the phenomenon in action.

Talk About It

Look at the photo and video of metal being moved. What questions do you have about the phenomenon? Talk about them with a partner. Record or illustrate your thoughts below.

Did You Know?

Pigeons have a magnetic sense that helps them find their home!

INQUIRY ACTIVITY

Hands On

Magnet Investigation

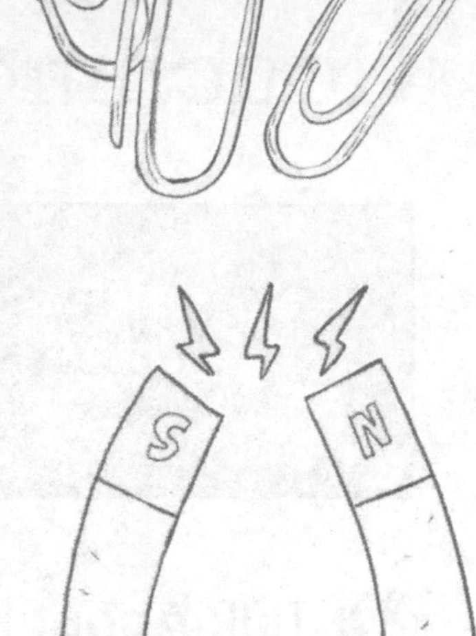

The video you watched showed a magnet lifting lots of metal. Magnets can move objects made of certain materials. You will investigate how magnets affect different materials.

Make a Prediction What objects can be picked up by a magnet? Write your predictions in the table below.

Carry Out an Investigation

1. Lay out the items you are going to test on a table.
2. Test each item by touching it with the magnet.
3. **Record Data** Record your observations in the table.

Will a Magnet Pick It Up?

Item	Predicted Results	Tested Results
penny		
plastic spoon		
paper clip		
pencil		
crayon		
aluminum foil		
metal spoon		

Materials

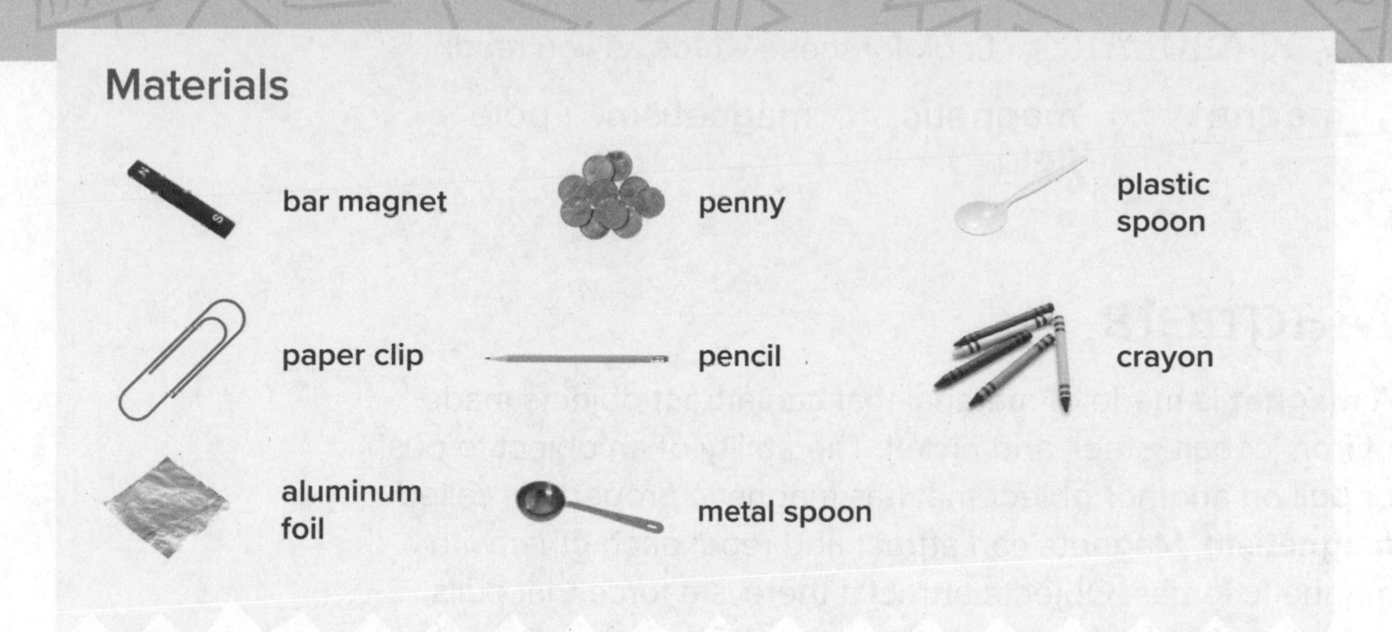

Communicate Information

4. Did your observations support your prediction? Explain.

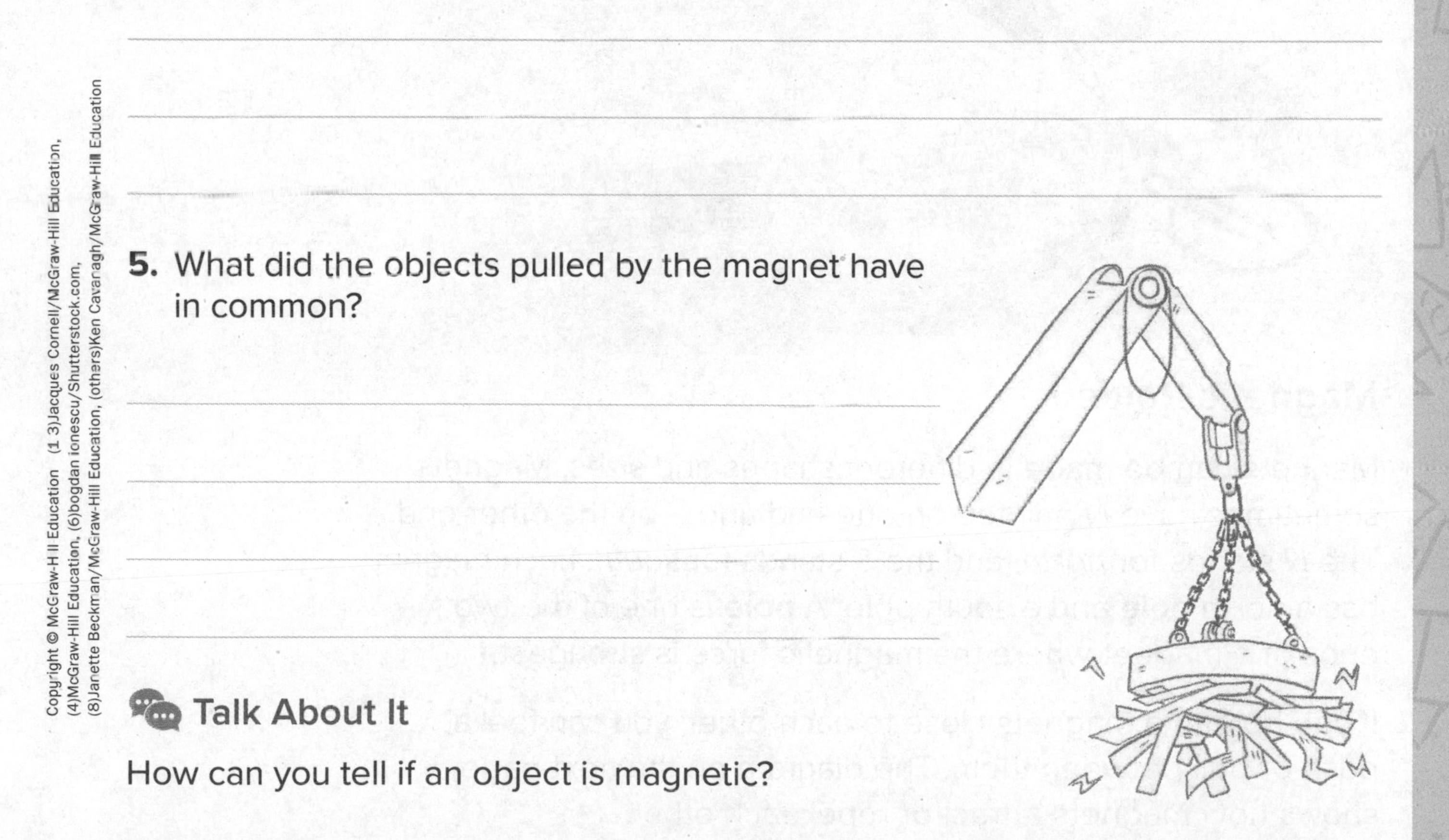

5. What did the objects pulled by the magnet have in common?

Talk About It

How can you tell if an object is magnetic?

VOCABULARY Look for these words as you read:

magnet | magnetic field | magnetism | pole

Magnets

A **magnet** is made of material that can attract objects made of iron, cobalt, steel, and nickel. The ability of an object to push or pull on another object that has magnetic property is called **magnetism**. Magnets can attract and repel each other with magnetic forces. Objects attract if there is a force that pulls them towards each other. Objects repel if there is a force that pushes them apart.

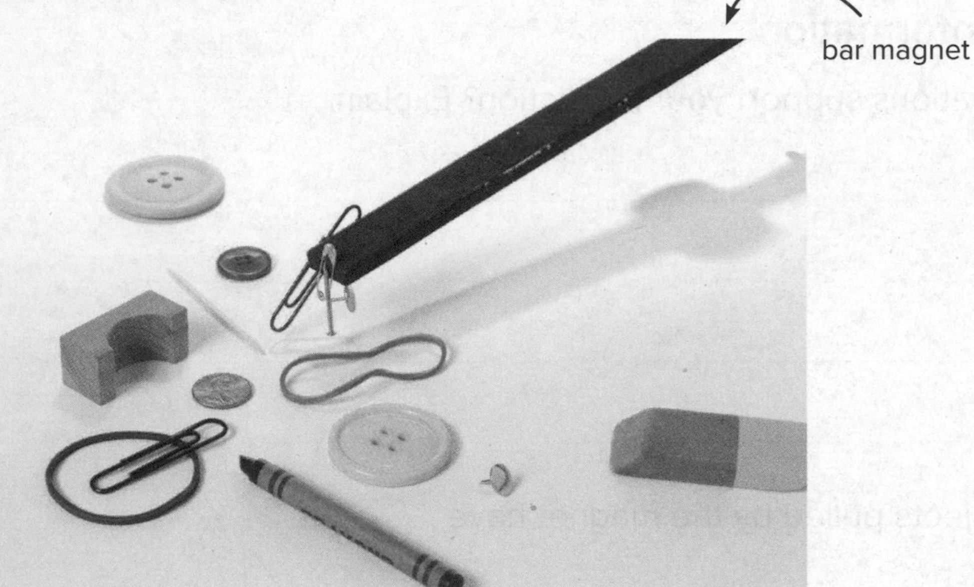

Magnetic Poles

Magnets can be made in different shapes and sizes. Magnets sometimes have *N* painted on one end and *S* on the other end. The *N* stands for *north,* and the *S* stands for *south*. Each magnet has a north pole and a south pole. A **pole** is one of the two ends of a magnet where the magnetic force is strongest.

If you hold two magnets close to each other, you can feel a push or pull between them. The diagram on the next page shows how magnets attract or repel each other.

Magnetic Field

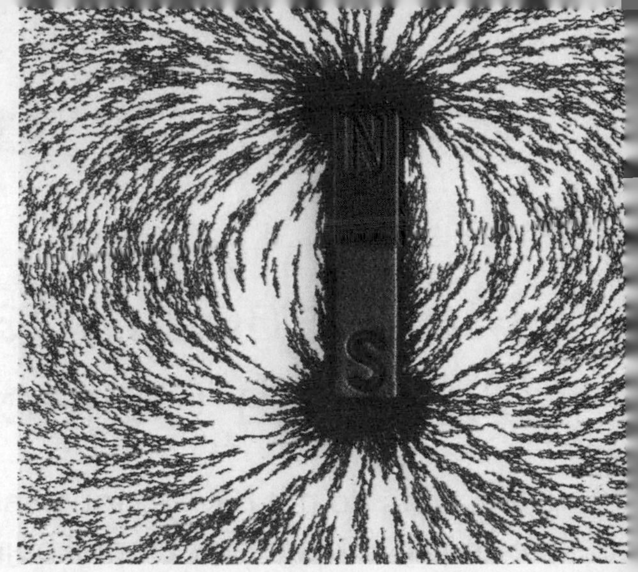

Bits of iron were sprinkled around this magnet. The bits of iron show the magnetic field.

If you want to throw a ball, your hand has to touch the ball. A magnet can push or pull an object without touching it. It does have to be close enough to the object to be in its magnetic field.

A **magnetic field** is the area around a magnet where its force can attract or repel. You cannot see a magnetic field, but you can feel where it is. If you bring two magnets close, you can feel them push or pull each other. Even when the magnets do not touch, their magnetic fields interact. If you move the magnets far apart, you do not feel the push or pull any longer. The magnetic fields are no longer meeting.

GO ONLINE Explore with *Effects of Magnets* to see the actions of a magnet's north and south poles.

Look back at the Inquiry Activity, *Magnet Investigation*. Why did the magnet only attract certain objects?

____

____

Label a Diagram

Label the two ends of the magnet with *N, north,* and *S, south* to represent the two ends of the magnets.

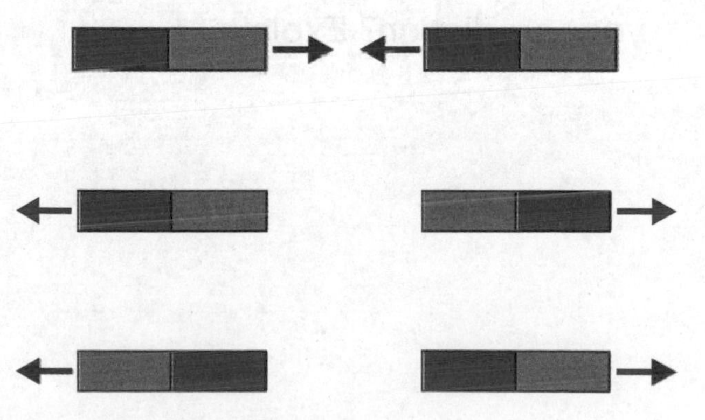

Two magnets attract each other when the south pole of one faces the north pole of the other.

Two magnets repel each other when their south poles face each other.

Two magnets also repel each other when their north poles face each other.

INQUIRY ACTIVITY

Hands On

Magnetic Forces Pass Through Objects

Magnets do not have to touch an object to attract or repel it. Magnets can pull objects through other materials.

Make a Prediction How will the thickness of the material a magnet is pulling through affect the strength of the pull?

Materials

bar magnet

paper clip

250 sheets of recycled paper

Carry Out an Investigation

1. Place a paper clip on top of a sheet of paper.
2. **Record Data** Observe how the magnet moves the paper clip. On a separate piece of paper, record your observations.
3. Repeat with different amounts of paper.

Communicate Information

4. Did the results of your investigation support your prediction? Explain.

Earth's Magnetic Field

You just learned that magnetic fields can pass through some objects and not others. Earth is a giant magnet. Iron deep inside Earth creates a huge magnetic field around the planet. Just like a bar magnet, Earth has two magnetic poles.

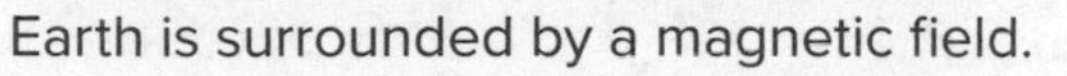

Earth is surrounded by a magnetic field.

A compass uses magnetism to show direction.

A compass is a tool that helps you find north, south, west, east, and other directions in between. The needle of a compass is a magnet that can move around. The red arrow always points north. Why? Earth's magnetic north pole attracts the compass needle. Before GPS was invented, people used compasses to determine where they were going.

Why does a compass arrow always point north?

INQUIRY ACTIVITY

Hands On

Make an Electromagnet

You have explored how magnets work to attract objects. In this activity, you will construct an electromagnet and consider how to make it stronger.

Make a Prediction How will the number of times you wrap a wire around a nail affect the strength of an electromagnet?

Carry Out an Investigation

BE CAREFUL Wire may be hot.

1. Starting at one end of the nail, wind the insulated wire around the nail 20 times. Leave at least 4 centimeters of wire at the starting end.
2. Place one of the D-cell batteries into one of the battery holders.
3. Attach each of the ends of the insulated wire into the clips on each end of the battery holder.
4. **Record Data** Use the nail as a magnet, and see how many paper clips you can pick up. Record the results in the table.
5. Wind the wire 10 more times around the nail. Record the results in the table.

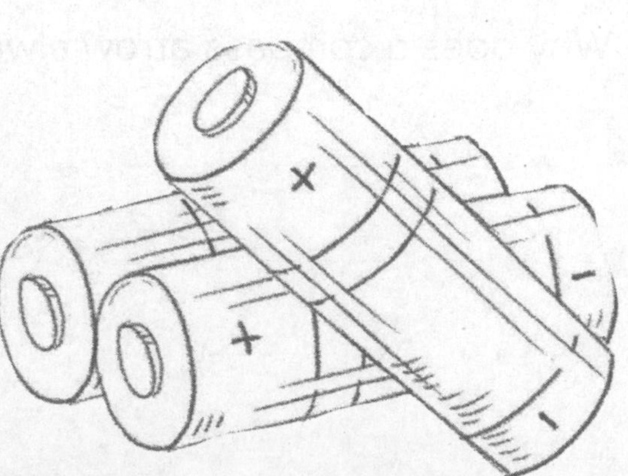

Strength of Electromagnet				
Number of Times Wire is Wound Around Nail	**1 Battery**	**Number of Paper Clips**	**2 Batteries**	**Number of Paper Clips**
20				
30				
40				

6. Wind the wire 10 more times. Record the results in the table. How did the number of times the wire was wound affect the number of paper clips picked up?

7. Now dissemble the electromagnet. Go back to the beginning of the activity. This time, connect a second battery in the series.

Communicate Information

8. Did your observations support your prediction? Explain.

9. How did adding a second battery affect the strength of the electromagnet?

Using Magnets

> **GO ONLINE** Watch the video *Magnets Solve Problems* about uses for magnets.

A magnetic field forms around a wire if a current flows in the wire. If you wind the wire into a coil, the field is stronger. When a current flows in the wire, the coil becomes a magnet. The magnet is stronger if you place a metal bar inside a coil. An electromagnet is a coil of wire around a metal bar, such as an iron nail. A battery at the ends of the wire makes a current flow in the wire.

You can turn an electromagnet on and off with a switch. The switch makes electromagnets useful in many electric devices, like speakers and doorbells.

1. What causes the magnetic field in an electromagnet?

____

____

2. What are two ways to make the electromagnet stronger?

____

____

____

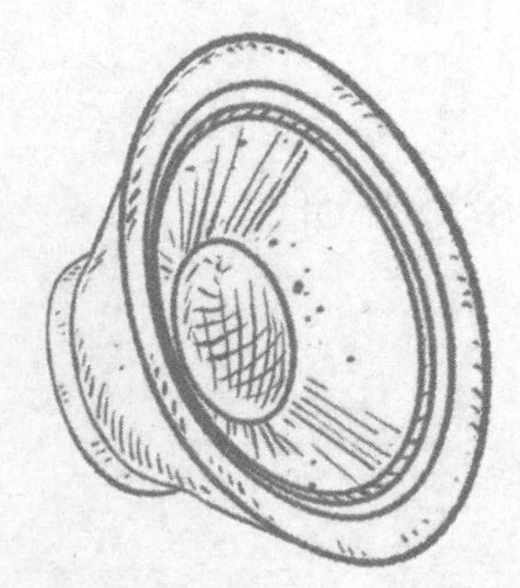

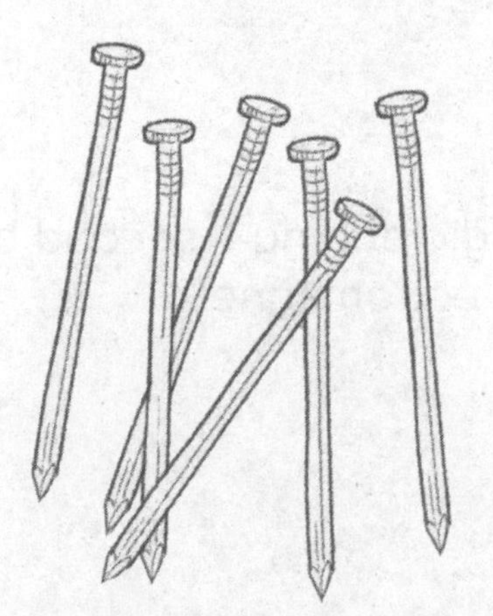

FOLDABLES®

Cut out the Notebook Foldables given to you by your teacher. Glue the anchor tabs as shown below. Describe the picture using vocabulary words.

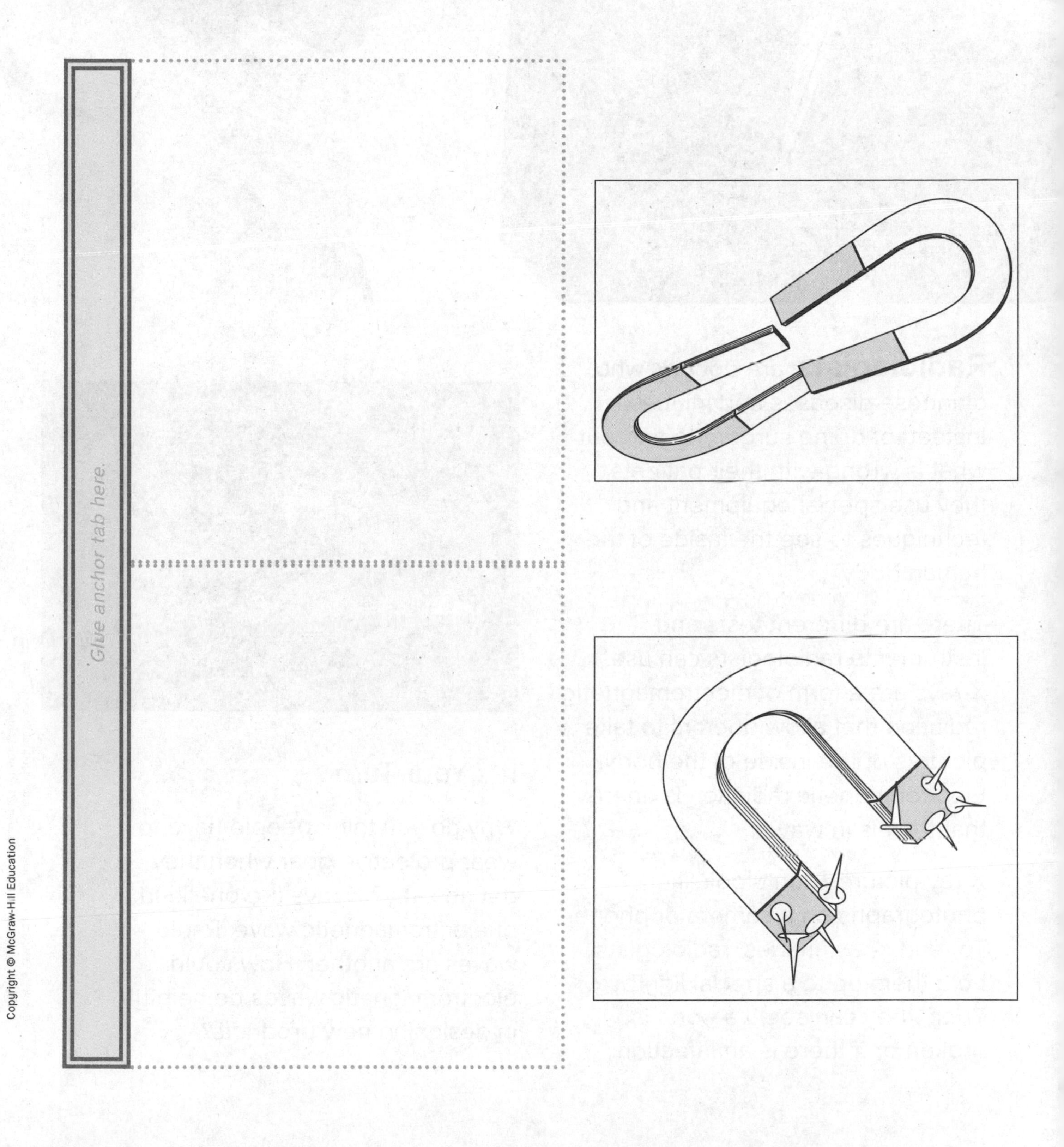

STEM Connection

What Does a Radiologist Do?

Radiologists are doctors who diagnose diseases and injuries. Instead of doing surgery to find out what is wrong with their patients, they use special equipment and techniques to see the inside of the human body.

There are different tests and instruments radiologists can use. X-rays are a form of electromagnetic radiation that allow doctors to take pictures of the inside of the body. Electromagnetic radiation is energy that travels in waves.

X-ray pictures don't look like photographs on a camera or phone. To read x-ray images, radiologists hold them up to a special lightbox. Then, they can see if a bone is broken or if there is an infection.

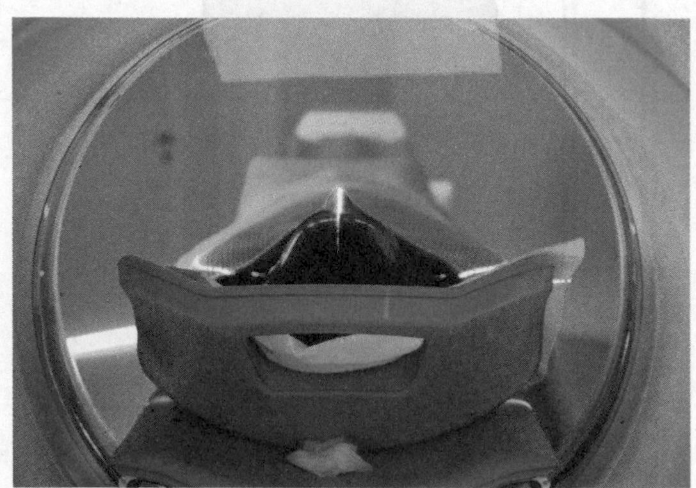

It's Your Turn

Why do you think people have to wear protective gear when they get an x-ray? X-rays are one kind of electromagnetic wave. Radio waves are another. How could electromagnetic waves be helpful in designing new products?

Light from Motion

A dynamo is a simple tool that turns motion energy into electricity (by spinning a magnet inside a coil of wire). A small dynamo moved by the wind, a water-wheel, or a bike can make enough electricity to power a light or a computer. A car battery can store the electricity for use later. You do have to buy a dynamo and battery, but after that, the electricity is free.

A dynamo can be used to power the bike's light.

WRITING Connection Magnetism has been observed for centuries. Around 600 BC, an Indian surgeon used magnets to remove metallic splinters. Around 1000 BC, the Chinese found they can use magnets to aid in navigation.

Read the Investigator article, *Medical Magnets*, to learn how magnets are used in medicine today. Write a paragraph about how magnets are used today, and then invent a new and unique use for magnets.

LESSON 2

Review

EXPLAIN THE PHENOMENON

How is that object being moved without anything touching it?

Summarize It

Explain how objects can be moved by a magnet without coming in contact with them.

REVISIT Revisit the Page Keeley Science Probe on page 67. Has your thinking changed? If so, explain how it has changed.

Three-Dimensional Thinking

1. What causes a magnet to attract and repel?

A. magnetic force

B. contact force

C. gravity

D. friction

2. Think of a task that you do every day at home. How do you think a magnet could improve the task?

____

____

____

Extend It

Imagine you are an electrical engineer. You are asked to design a device that uses magnetism and electricity to clean up litter in a park.

Draw your plan for the device. Then, explain how your device will use magnetism and electricity to solve the litter problem.

KEEP PLANNING

STEM Module Project
Engineering Challenge

Now that you have learned how magnets can move objects, go to your Module Project to explain how the information will affect your plan for your gate.

STEM Module Project
Engineering Challenge

Design a Self-Closing Gate

You've been hired as an engineer. Using what you have learned throughout this module, you will develop a design for a self-closing gate. Your goal will be to design and select the appropriate materials for a gate that can close itself.

Planning after Lesson 1

Apply what you have learned about electricity and designing solutions to your project planning.

How does knowing about electricity affect your module planning?

Record information to help you plan your model after each lesson.

Planning after Lesson 2

Apply what you have learned about magnetism and designing solutions to your project planning.

What properties of magnets can help you with your module project?

Research the Problem

Research ideas for materials that could be used in your project by going online to teacher-approved websites, by interviewing an engineer, or by finding a book about gate closures from your local library.

Source	Information to Use in My Project

Sketch Your Own Design

Draw your ideas. Select the best one to build and test.

STEM Module Project
Engineering Challenge

Design a Self-Closing Gate

Look back at the planning you did after each lesson.
Use that information to complete your final module project.

The Engineering Design Process

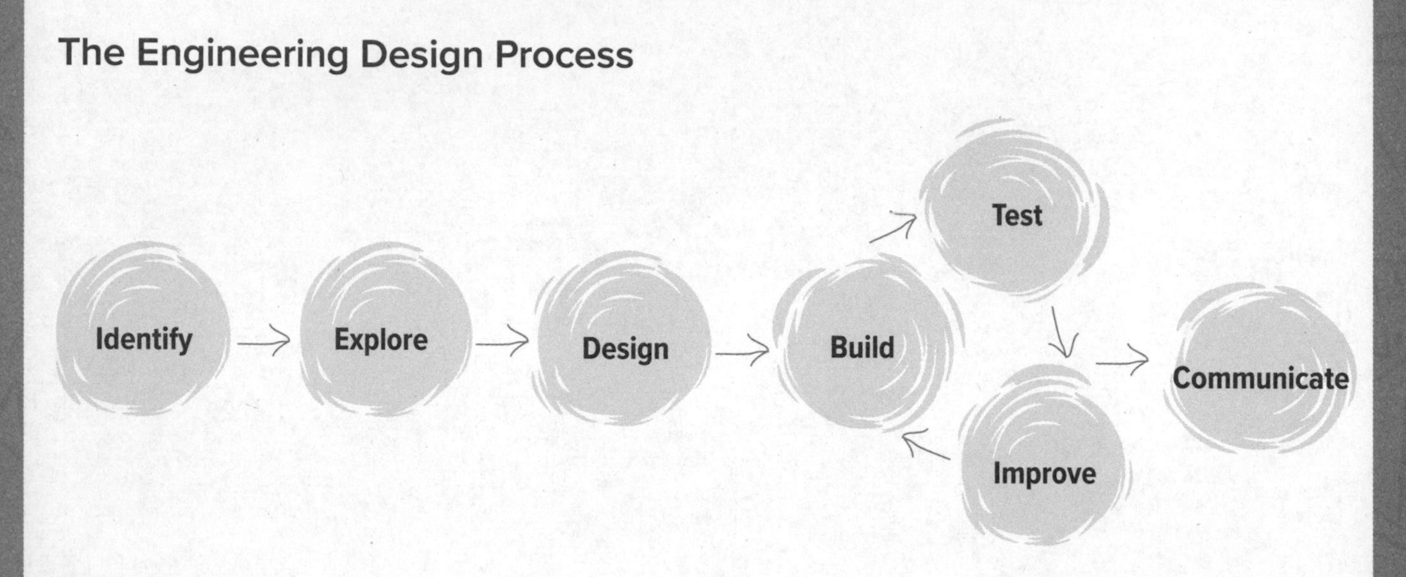

Build Your Model

1. Use your project planning.
2. Write clear steps to build and test your model.
3. List the materials you will use in the space provided.
4. Draw your design on the next page. Follow your procedure to build and test your design.
5. Record the improvements made to the design after each trial.

Materials

Procedure:

Test Your Model

Build and test your model. Record your observations and results. Use a data table if you need to.

You are using the Engineering Design Process!

STEM Module Project
Engineering Challenge

Communicate Your Results

Share and compare the plan for your model and your results with another group. Communicate your findings below.

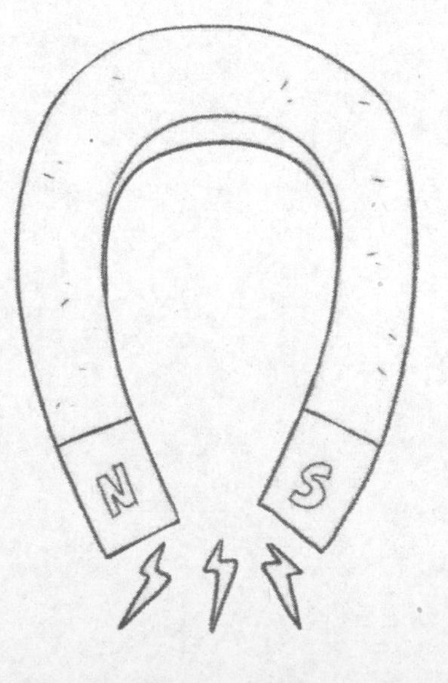

MODULE WRAP-UP

REVISIT THE PHENOMENON

Using what you learned in this module, explain how electricity and magnets can be used to move objects.

Have your ideas changed?

Science Glossary

A

adaptation a structure or behavior that helps an organism survive in its environment

atmosphere a blanket of gases and tiny bits of dust that surround Earth

attract to pull toward

axis an imaginary line through Earth from the North Pole to the South Pole

B

balanced forces forces that cancel each other out when acting together on an object

birth the beginning or origin of a plant or animal

C

camouflage an adaptation that allows an organism to blend into its environment

climate the pattern of weather at a certain place over a long period of time

competition the struggle among organisms for water, food, or other resources

D

direction the path on which something is moving

distance how far one object or place is from another

E

ecosystem the living and nonliving things that interact in an environment

electrical charge the property of matter that causes electricity

environmental trait a trait that is affected by the environment

extinction the death of all of one type of living thing

F

floodwall a wall built to reduce or prevent flooding in an area

force a push or pull

fossil the trace of remains of living thing that died long ago

friction a force between two moving objects that slows them down

G

germinate to begin to grow from a seed to a young plant

group a number of living things having some natural relationship

H

hibernation to rest or go into a deep sleep through the cold winter

I

inherited trait a trait that can be passed from parents to offspring

instinct a way of acting that an animal does not have to learn

invasive species an organism that is introduced into a new ecosystem

L

learned trait a new skill gained over time

levee a wall built along the sides of rivers and other bodies of water to prevent them from overflowing

life cycle how a certain kind of organism grows and reproduces

lightning rod a metal bar that safely directs lightning into the ground

M

magnet an object that can attract objects made of iron, cobalt, steel, and nickel

magnetic field the area around a magnet where its force can attract or repel

magnetism the ability of an object to push or pull on another object that has the magnetic property

metamorphosis the process in which an animal changes shape

migrate to move from one place to another

mimicry an adaptation in which one kind of organism looks like another kind in color and shape

motion a change in an object's position

N

natural hazard a natural event such as a flood, earthquake, or hurricane that causes great damage

P

pole one of two ends of a magnet where the magnetic force is strongest

pollination the transfer of pollen from the male parts of one flower to the female parts of another flower

population all the members of a group of one type of organism in the same place

position the location of an object

precipitation water that falls to the ground from the atmosphere

R

repel to push away

reproduce to make more of their own kind

resource a material or object that a living thing uses to survive

S

season one of the four parts of the year with different weather patterns

static electricity the build up of an electrical charge on a material

survive to stay alive

speed a measure of how fast or slow an object moves

T

temperature a measure of how hot or cold something is

trait a feature of a living thing

U

unbalanced forces forces that do not cancel each other out and that cause an object to change its motion

V

variation an inherited trait that makes an individual different from other members of the same family

W

weather what the air is like at a certain time and place

Index

T

U

W

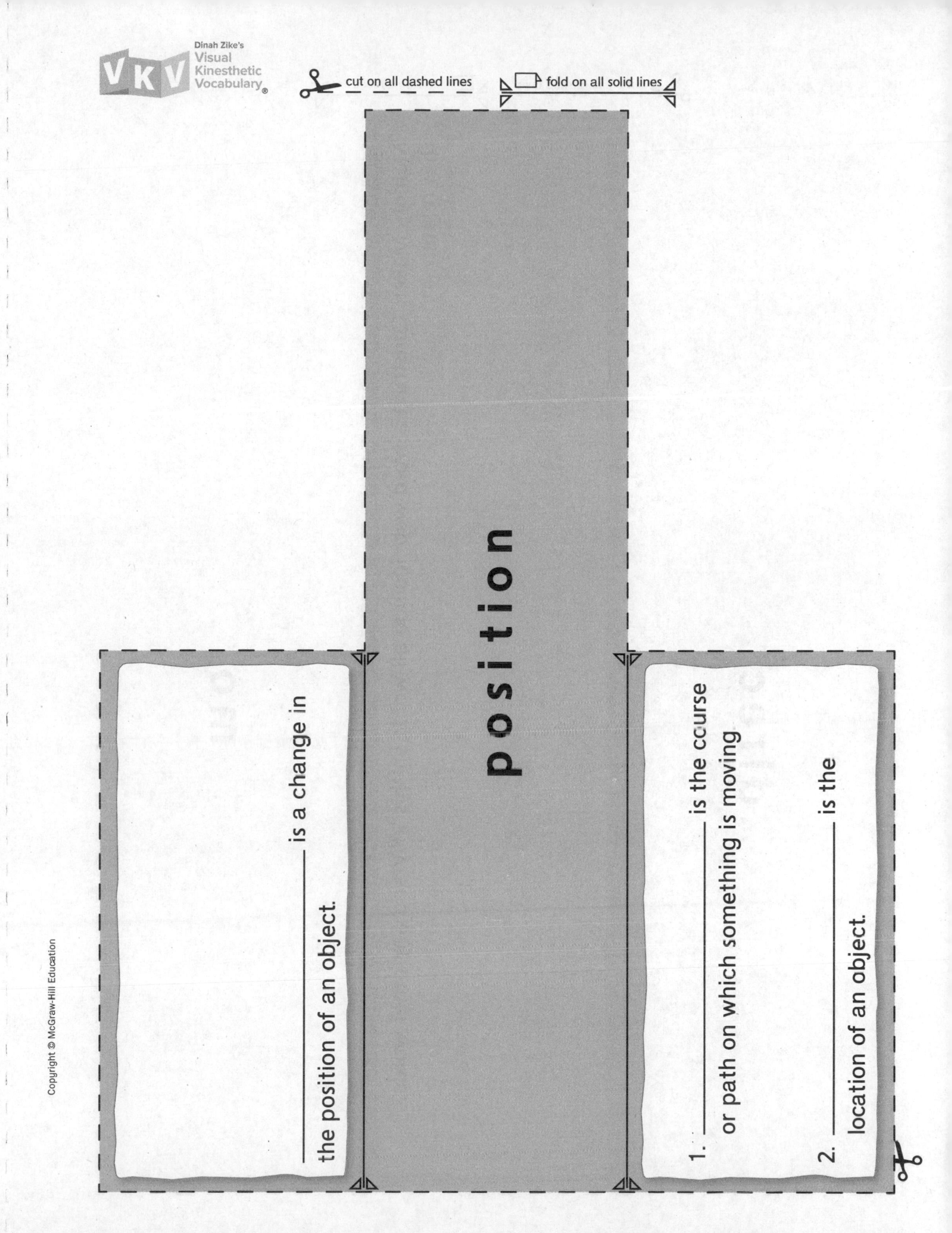
Dinah Zike's
Visual Kinesthetic Vocabulary®
cut on all dashed lines
fold on all solid lines
position
______ is a change in the position of an object.
1. ______ is the course or path on which something is moving.
2. ______ is the location of an object.

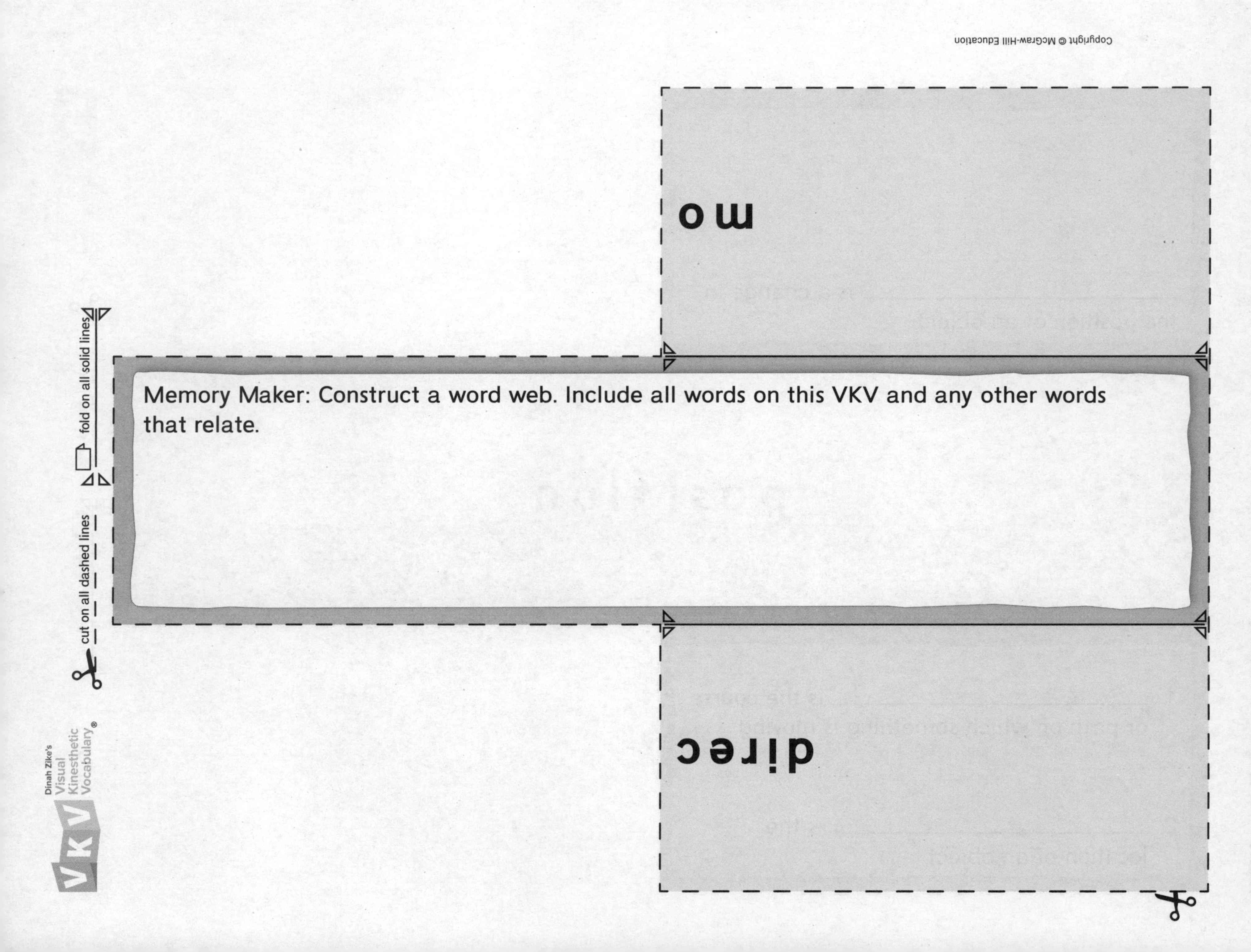

Dinah Zike's
VKV
Visual
Kinesthetic
Vocabulary®
cut on all dashed lines
fold on all solid lines
Memory Maker: Construct a word web. Include all words on this VKV and any other words that relate.
mo
direc

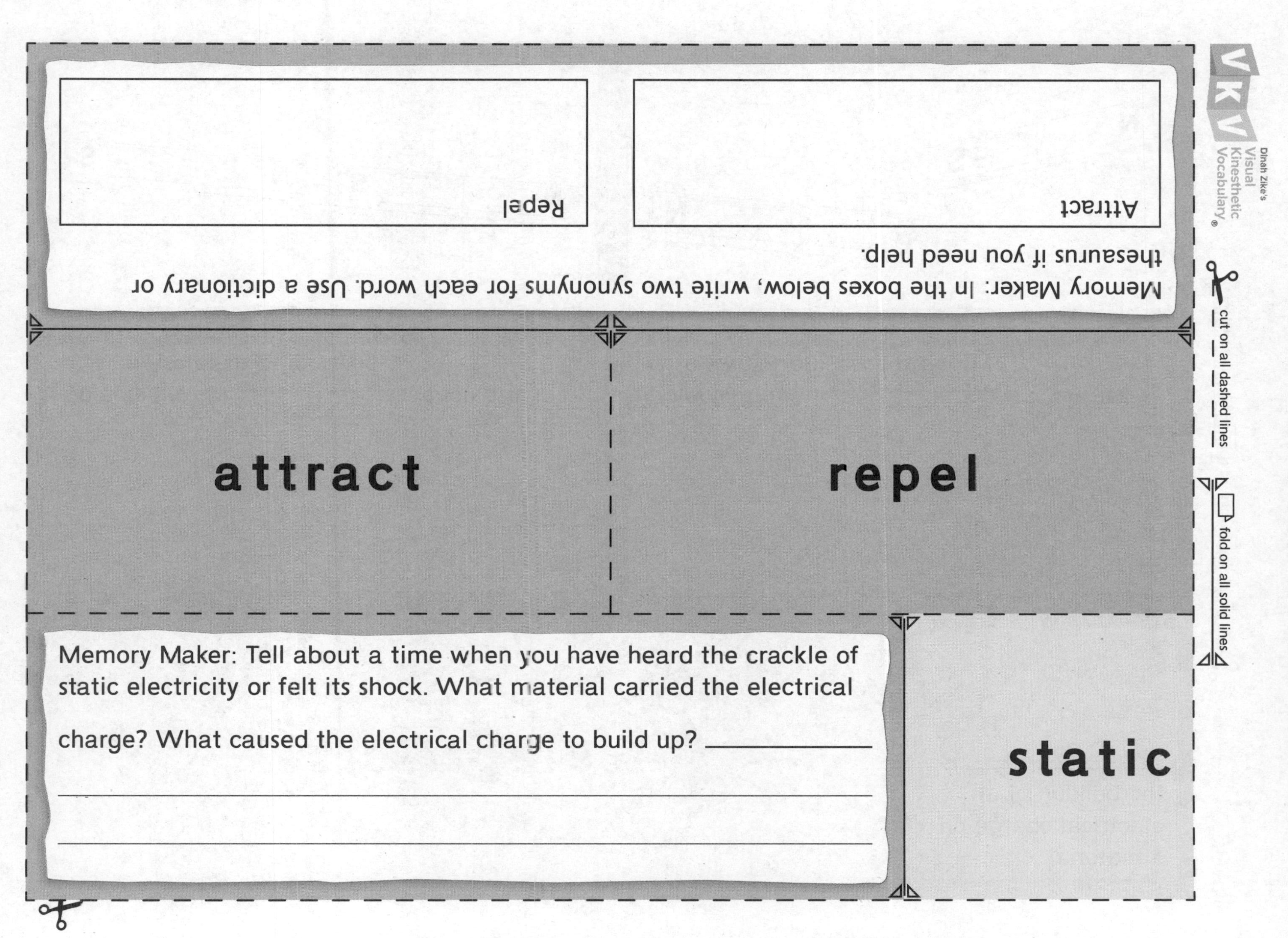
Dinah Zike's Visual Kinesthetic Vocabulary®
cut on all dashed lines
fold on all solid lines
Memory Maker: In the boxes below, write two synonyms for each word. Use a dictionary or thesaurus if you need help.
Attract
Repel
attract
repel
Memory Maker: Tell about a time when you have heard the crackle of static electricity or felt its shock. What material carried the electrical charge? What caused the electrical charge to build up?
static

Dinah Zike's Visual Kinesthetic Vocabulary®

cut on all dashed lines

fold on all solid lines

______ is the buildup of an electrical charge on a material.

electricity

Objects that ______ each other push away from each other.

N S S N

Objects that ______ each other pull toward each other.

N S N S